Nature's Living Lights

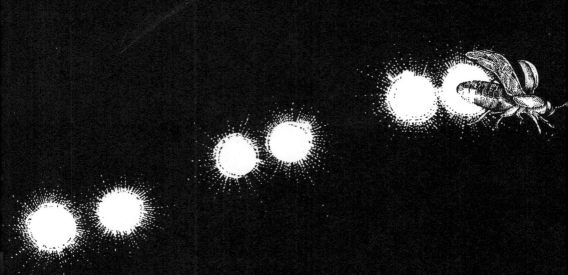

NATURE'S LIVING L·I·G·H·T·S

Fireflies and Other Bioluminescent Creatures

by Alvin and Virginia Silverstein

Illustrated by Pamela and Walter Carroll

LITTLE, BROWN AND COMPANY   BOSTON   •   TORONTO

FIRST EDITION

Library of Congress Cataloging-in-Publication Data

Silverstein, Alvin.
    Nature's living lights.

    Summary: Describes bioluminescent insects, plants
and sea animals and the uses these creatures make of
their self-generated lights, such as communication,
finding food, and attracting mates.
    1. Bioluminescence — Juvenile literature.
[1. Bioluminescence]   I. Silverstein, Virginia B.
II. Title.
QH641.S54   1988   574.19'125   87-2727
ISBN 0-316-79119-9

10   9   8   7   6   5   4   3   2   1
DESIGNED BY JEANNE ABBOUD

BP
*Published simultaneously in Canada
by Little, Brown & Company (Canada) Limited*

PRINTED IN THE UNITED STATES OF AMERICA

For Gupi Elinon Silverstein

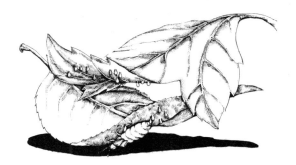

Nature's Living Lights

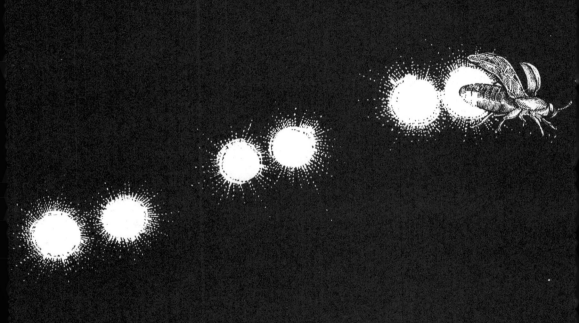

ON a warm summer night, tiny lights wink on and off with a bright glow. Sometimes the air seems full of them.

These lights are carried by little flying creatures. Put out your hand, and one may land on your finger. It looks like an ordinary beetle until the end of its body lights up like a tiny lantern. It is a firefly.

Fireflies are not actually flies; they are beetles. There is something not quite right about the other part of their name, either. A fire gives off bright light, but it is also very hot. If you touch a firefly's lighted "lantern," you will find that it is not hot at all. This little beetle makes a cold light. Chemical reactions inside the firefly's body produce energy that is nearly 100 percent light and only about 1 percent heat. Even the best electric light bulbs are not nearly that good. Less than a quarter of their energy comes out as light, and the rest is all wasted as heat. Scientists call the firefly's light *bioluminescence* — light made by living things.

A Living World of Lights

MANY animals and plants are bioluminescent. Rotting meat or fish may glow with an eerie cold light. The light is produced by bacteria, tiny creatures that cannot be seen without a microscope. The amount of light from a single bacterium is very small — so small that it would take 100,000,000,000,000,000,000 of them to equal a sixty-watt light bulb. But bacteria multiply and grow in huge numbers. At the International Exhibition in Paris in 1900, a scientist who was studying luminous bacteria filled jars with them and placed them around the room. Their light was so bright that people could read by it and even recognize the faces of friends standing twenty feet away.

Luminous bacteria live in the sea, too. So do tiny creatures called *dinoflagellates*. They are bigger than bacteria but still too small for you to see. Under a microscope, though, you can see that each one is covered with a tiny shell, and they have interesting and beautiful shapes. These tiny creatures float in the water near the surface. When the water is quiet, they are dark. But if a ship or a fish comes by, the water suddenly glows with a flashing, sparkling "fire."

Many sea animals produce bioluminescence. Luminous jellyfish float on the surface like glowing dinner plates. There are shrimps and clams that carry their own lights with them. *Cypridina*, a clamlike creature about the size of a tomato seed, can shoot a bright blue cloud out into the water. If the bodies of Cypridinas are dried and then mixed with water, they produce a blue glow. They can still glow even if they have been kept dried for twenty years. During World War II, Japanese soldiers carried Cypridina powder with them. Sometimes they put a little moistened Cypridina powder on the back of their uniform so that their friends could see where they were as they crept through the jungle. If a soldier wanted to read a map, he mixed a little Cypridina powder with water in the palm of his hand. The dim blue glow gave him just enough light to read by.

At the bottom of the ocean the water is always dark. It is so deep that the light of the sun cannot reach it. Most of the fish and other animals that live deep in the ocean carry their own lights with them. Some carry dangling lanterns. Others have bright "headlights" under their eyes or on their bellies. Some have rows of lights along their sides that look like the lighted portholes of an ocean liner.

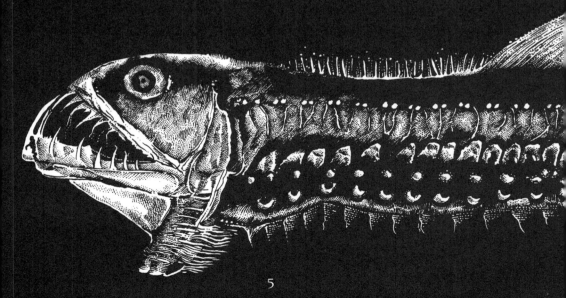

How Living Lights Glow

ALL the bioluminescent creatures, whether they live on the land or in the sea, make their living lights in the same way. They use a chemical reaction. Four things are needed for this reaction. First there must be two special light-producing chemicals made in the creature's body. Scientists have named these chemicals *luciferin* and *luciferase*. (*Lucifer* means light carrier.) When luciferin and luciferase are mixed together, they produce light. But they won't react unless two other chemicals are there, too. One is a sort of energy package called *ATP*. All living things use ATP for energy. Every time you blink an eye or move your hand, your movements are powered by ATP. The last ingredient needed for the reaction that makes living lights is something else that all living things use: *oxygen*. This is a gas that is part of the air we breathe. When oxygen joins with the complicated chemicals of living things, it helps to let out some of their stored energy. When a fire burns, oxygen joins with chemicals to let out energy in the form of heat and light. But in the reactions of bioluminescence, nearly all the energy that comes out is light.

If you mix some luciferin from a firefly with a bit of firefly luciferase in a test tube, and add some ATP and oxygen, the chemicals will glow. But if you mix firefly luciferin with luciferase from Cypridina, nothing happens. Each kind of bioluminescent creature makes its own light-producing chemicals. But there are some exceptions. For example, the luciferin and luciferase made

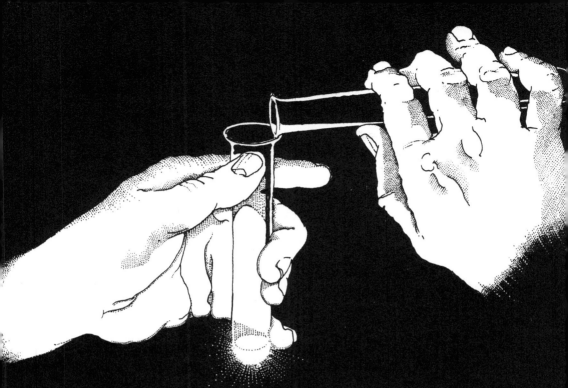

by certain luminous bacteria are exactly the same as those found in the "living lanterns" of some luminous fish. Scientists have discovered that there is a good reason for this. These fish don't make their own light. They eat the bacteria, and then some of them grow inside the fish's body. It is a good partnership for both of them. The fish provide a home for the bacteria and give them all the food and oxygen that they need. And the bacteria provide lights for the fish that swim in the dark ocean waters.

Fireflies, on the other hand, do make their own light. There are about two thousand different kinds of fireflies. Not all of them light up when they are adults, but they all glow when they are young. The young firefly beetle, called a *larva*, looks rather like a worm. So the glowing firefly larvae are often called glowworms.

A Firefly's Life

THE fireflies of the eastern United States start life as tiny eggs stuck to the underside of a leaf. A female firefly lays about a hundred eggs, which hatch in about four weeks into wiggly little larvae. Each one is less than an eighth of an inch long, but they grow fast. They creep along the soil, eating small soil insects, snails, and earthworms. The firefly larva has a poisonous bite. Its poison kills its prey and turns its insides to liquid. Then the larva sucks up its food like soup. In the fall, when the frost comes, the larva crawls under a stone and goes to sleep until spring. During the warm months, it crawls around and feeds and grows again. But in the fall it settles down for another winter sleep. The following

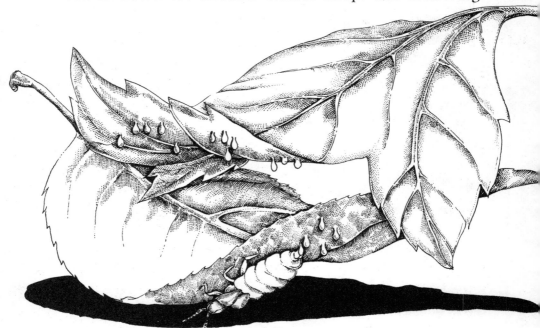

spring, it feeds for only a couple of months. Late in May it digs itself a shallow pit and chews up soil to make a roof from strips of mud. Hidden inside, the plump larva becomes a *pupa*. It does not move, but within its tough outer covering, amazing changes are taking place. The larva seems to melt. Its old eyes and legs and even its glowing lamps disappear, and a new creature grows in its place. It looks very different. Its eyes are much bigger. It has long insect legs instead of the short, stubby larval legs. And it has wings. It does not look like a worm anymore; it looks like a beetle. All these changes take about ten days. Then the adult firefly is ready to break through the roof of its hiding place and come out. Now it has just one big job left to do: to find a mate and start the next generation of fireflies.

10

Flashing the Mating Game

THE fireflies' lights now come in handy. Males and females use the lights to signal to each other. At twilight in the early summer the male fireflies fly out over the fields and through the woods, blinking their lights on and off in a regular pattern. The male is a strong flier, and when he turns on his light the whole back end of his abdomen shines brightly. His blinking light is like an announcement: "Here I am, a male firefly looking for a mate." Down on the ground or perched on stalks of grass, females are waiting. They are looking for mates too. When a female sees a flying light, blinking in just the right pattern, she waits about two seconds and then blinks back. Her light is much smaller, but the male's keen eyes can spot it. He flashes out his message again, and the female answers. After a time or two more, he flies down to meet her.

Different kinds of fireflies have different patterns of flashing lights. The most common type of firefly in the eastern United States, called *Photinus pyralis*, has a yellow light that flashes every six or seven seconds. Each flash lasts about half a second. While his light is on, the male swoops down and up again, so that he makes a glowing J-shape in the air. When the weather turns cool, the firefly takes longer between flashes. Chemical reactions slow down when it is cool and speed up when it is hot, and the reactions of bioluminescence are no exception. Somehow the female firefly can still recog-

nize the slower blinking pattern when the weather is cooler. She answers more slowly, too.

Other kinds of fireflies have different flashing patterns. Some flash quickly twice in a row, then stop, then flash quickly twice again. Some have a pattern with even more flashes — as many as eleven quick blinks of light. Having a different pattern for each kind of firefly means that each one can find the right kind of mate.

Lying Lights

THERE is another common kind of firefly that doesn't play by the rules. This is a kind called *Photurus*. (The "Phot-" part of both Photinus and Photurus comes from a word meaning light. Photograph, a picture made with light, comes from the same word.) Photurus males and females have their own flashing pattern, using a green-ish light. But after a Photurus female has mated, she starts to cheat. She answers not only Photurus males, but also males of other kinds. Each time, she uses the answering pattern of their own species, pretending that she is the right kind of mate for them. She can imitate as many as seven other flashing patterns. If a male is fooled and flies down to join her, he gets an unpleasant surprise. She pounces on him and eats him! She may even fly up after a male and catch him in the air. Most female fireflies are not very good fliers, but Photurus is an exception. She is very strong. If you catch a Photurus female in your hand, she will wriggle, brightly flashing her light. She may even bite.

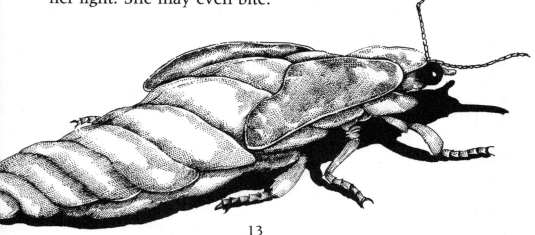

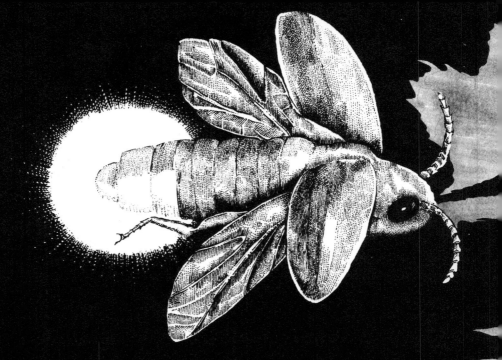

With cheaters like Photurus around, things get very complicated in the world of living lights. Some male fireflies have learned to protect themselves. Instead of flying straight down to an answering female, they immediately drop out of the air when they see an answering light. Then they walk up to the female on foot — very carefully. Photurus males looking for a mate sometimes imitate the flashing patterns of males of other species — the ones that Photurus females want to eat. The Photurus male flashes his own pattern while he is flying high up over the trees. But then as he starts to fly down, he switches to another pattern, and then another. If he sees an answering flash, he flies down to investigate. The answering female may be a Photurus pretending to be the kind of firefly that the male is pretending to be. Then he will try to persuade her to mate with him instead of eating him.

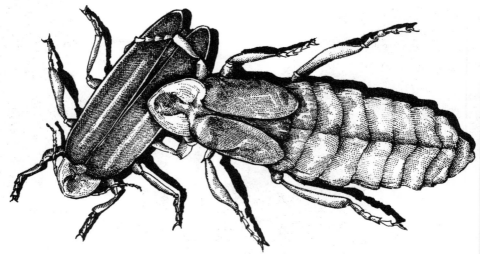

The cheating Photurus females have found themselves a very good food supply. On a summer evening there are a lot of male fireflies out looking for mates and only a few females to answer them. One scientist watched about two hundred male fireflies flying over a distance of ten miles one night. They flashed a total of nearly eight thousand times, but they found only two females to answer them.

The fireflies that live in the eastern United States are sometimes called lightning bugs because their lights flicker on and off like a flash of lightning. Messages carried along the firefly's nerves to its abdomen cause luciferin and luciferase to mix and turn on the light. But there are some fireflies that keep their lights on all the time in a steady glow. Scientists used to wonder why none of the glowing fireflies lives in the eastern United States, only the flashing ones. Now they think the answer is Photurus. A male firefly flying through the air with a steady glow would make too good a target for Photurus females. A light that blinks on and off is harder to hit, and so some of the flashing males escape and mate.

Living Lanterns

ONE kind of firefly with a steady glow is the fire beetle, which lives in the Caribbean and the tropics of South America. Its scientific name is *Pyrophorus*, which means fire carrier. It belongs to the family of click beetles. Fire beetles have two bright greenish lights, one on each shoulder, like the headlights of a car. Some kinds also have a third light on the underside of their bodies, which can be seen only when they are flying.

The natives call fire beetles *cucujos*, and they use them as living lanterns. They keep them in cages to light their houses. (About forty cucujos make as much light as a sixty-watt bulb.) People tie these glowing beetles to their hands and feet to light their way through the jungle. And women put them in mesh bags and use them as glowing hair ornaments.

Fireflies that live in southeast Asia put on a fantastic show. Males gather in trees in such huge numbers that every leaf is covered with them. Then they begin to flash on and off *all together*. This is called synchronized flashing. (Among the fireflies of the United States, each male does his own thing, and it is rare to see two or more blinking on and off at the same time.) A tree full of the Asian fireflies looks like a giant Christmas tree with blinking lights. Flashing together helps the fireflies to find mates. Trees and plants in the jungle grow so thickly that a female might not be able to see the flash of a single male. But the brilliant light shows of thousands of males all flashing together attract swarms of

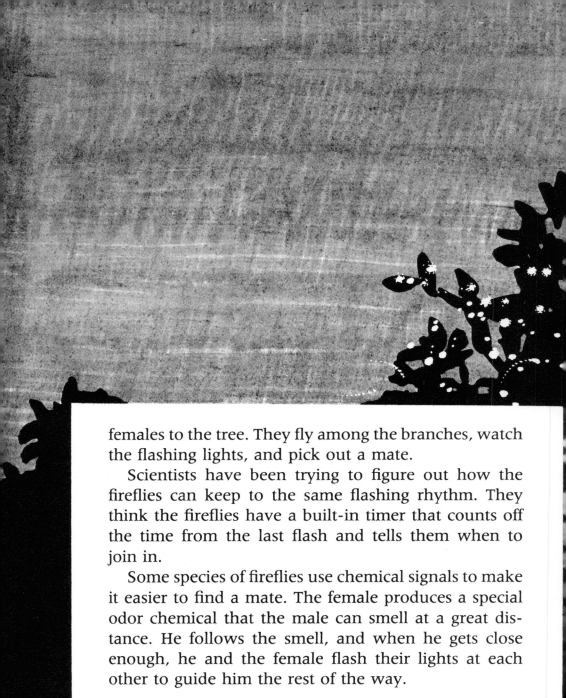

females to the tree. They fly among the branches, watch the flashing lights, and pick out a mate.

Scientists have been trying to figure out how the fireflies can keep to the same flashing rhythm. They think the fireflies have a built-in timer that counts off the time from the last flash and tells them when to join in.

Some species of fireflies use chemical signals to make it easier to find a mate. The female produces a special odor chemical that the male can smell at a great distance. He follows the smell, and when he gets close enough, he and the female flash their lights at each other to guide him the rest of the way.

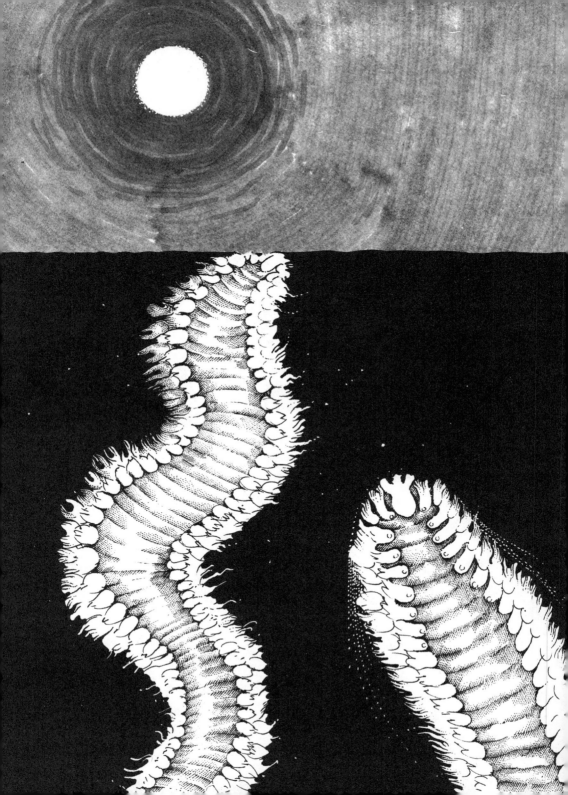

Lights in the Sea

FIREFLIES are not the only animals who use living lights to find a mate. Sea creatures also use light signals in this way. In the waters of the West Indies, glowing worms called *fireworms* produce an amazing light show a few times a year. These worms live quietly in burrows at the bottom of the sea. But two nights after each full moon, the fireworms suddenly leave their burrows and swarm up to the surface. The females, which shine all over with a steady glow, swim up first. Then the males, each with two small headlights, join them. The males and females swim around together in circles, flashing and glowing. The females send out a glowing cloud of eggs into the water, and the males send out sperm. The parents dive back to their burrows, while on the surface the sperm and eggs join to start the lives of new baby worms.

Finding a mate is just one of the uses for living lights. Bioluminescence can also help an animal to find food and escape from its enemies. In the dark waters at the bottom of the ocean, living lights may be the only way for animals to find their way around. And they also make good signals for "talking" to other members of their species.

The *flashlight fish*, which lives in the Indian Ocean and Red Sea, uses bioluminescence for all these things. This fish carries living luminous bacteria in special pockets called *light organs*, just below its eyes. It treats its little guests very well, providing food and oxygen for them and taking away their waste products. Scientists have tried to grow the bacteria in the laboratory, but usually they die within a week. Yet these same bacteria can go on living and multiplying in the light organ of a fish for twenty years.

As the flashlight fish swims, its two headlights shine out like the beam of a car's headlights. In fact, French divers who saw the fish called it the little Peugeot. Its scientific name, *Sparus palpebratus*, means the porgy with an eyelid. The fish can turn off its light by covering each light organ with a special "eyelid." Cream-colored on the outside and black on the inside, this lid blocks out the light completely when it is closed.

The flashlight fish's headlights help it to see where it is going and also help it to see its prey. They probably also help to bring prey to the fish, because many tiny sea creatures are attracted to light. Scientists think the fish also use their lights to "talk" to each other. In the laboratory, when two flashlight fish are placed in tanks next to each other, they blink their lights much faster than they do when they are alone. In their home in the sea, male and female fish blink back and forth at each other when they are choosing a mate. The fish use light to defend their home territory. When another fish swims near, a female flashlight fish turns off her lights, swims

up close to the stranger-fish, then suddenly turns on her lights in a startling flash. When a flashlight fish is trying to escape from a bigger fish, it uses a "blink and run" trick. It swims in zigzags and blinks its light on and off rapidly to confuse its enemy.

Some sea creatures use bioluminescence in different ways to escape from enemies. Some fish use light to become "invisible." In shallow waters and midwaters, sunlight comes down from the surface. From below, a swimming fish is clearly outlined as a dark shape against a lighter background. Some fish solve this problem by having glowing lights on their bellies. They can adjust the amount of light that is showing, so that it is just enough to match the amount of sunlight shining through the water. They turn their lights off at night and on in the daytime so that they always blend into the background.

Squids protect themselves by sending lights out into the water. Squids that live in shallow water, where the sunlight is bright, send out a cloud of black ink when they are frightened. Then they slip away while their enemy is blinded by the blackness. Black ink wouldn't be much help down at the bottom of the ocean, where everything is black anyway. Yet the squids that live on the ocean bottom use the same kind of trick, but instead of black ink, they shoot out luciferin and luciferase. The chemicals mix in the water and form a cloud of blinding brightness. Remember tiny, clamlike Cypridina that Japanese soldiers carried around for a reading light? Living Cypridina use their glowing blue light in the same way as squids, to dazzle the eyes of predators.

The *brittle star,* a relative of the starfish, uses lights to protect itself in an even stranger way. The brittle star lives in a burrow on the ocean bottom. Its long, thin arms stick up out of the burrow. If a predator pokes at the arms, they flash. If that doesn't scare the predator away, one of the arms breaks off and wriggles away like a snake, glowing brightly. Meanwhile the rest of the brittle star goes dark and hides in its burrow. It can grow a new arm to replace the one it lost.

Fright Lights on Land

ON land, various insects flash lights to frighten or confuse their enemies. One of the most unusual is the *railroad worm* of Central and South America. It is the larva of a beetle, and it grows to about an inch and a half long. It has two red headlights that shine when it is resting. (The railroad worm is one of the few insects that produces red light.) When the larva is disturbed, eleven pairs of green lights glow along its sides, like the lighted windows of a railroad train.

Another beetle has two glowing spots that look like a pair of glaring eyes. Other creatures seem to be frightened by these fiery "eyes" and leave this beetle alone. But the beetle is really fooling them. Those fierce-looking "eyes" are only shining spots on its body.

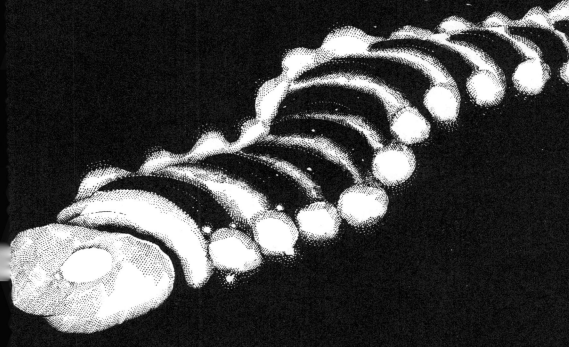

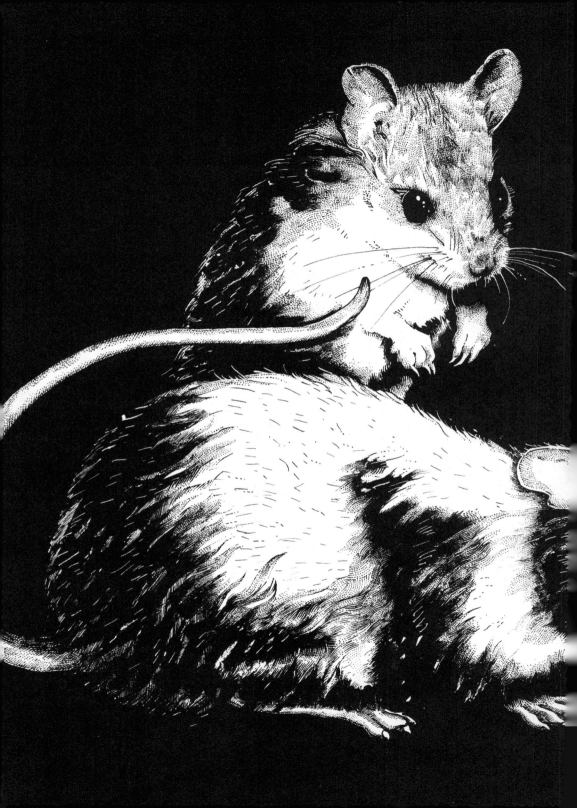

Scientists believe that the glow of *glowworms*, the larvae of fireflies, is a kind of advertisement telling other animals that they are not good to eat. Firefly larvae's bodies produce some very bad-tasting, poisonous chemicals called *lucibufagins*. Birds refuse to eat them, and so do some other animals. One group of researchers tried feeding glowworms to mice. The mice wouldn't eat them, so the researchers hid the larvae inside gelatin capsules. As soon as the mice bit into the larvae, they spit them right out. Then they spent the next few minutes wiping at their mouths with their forefeet. Tasting bad is not a very good defense if a predator doesn't find out about it until after it has eaten the animal. But tasting bad and being easy to recognize make a combination that helps at least some members of a species to survive. Once a mouse has tasted a firefly larva, it will remember its lesson. It will not be likely to bite into any more creatures that glow.

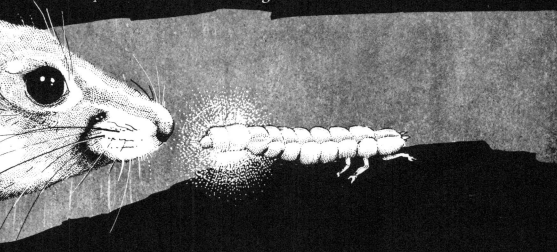

Light Lures

ANOTHER kind of glowworm uses bioluminescence for a very different purpose. While the glowworms of Europe and America use light to help them avoid being eaten, the glowworms that live in New Zealand use living lights to help them catch their food. These glowworms are not beetle larvae. They are the larvae of flies. (But they don't grow up to be fireflies. The adults look like mosquitoes and don't glow.) The larvae live in caves and shady places. They live in clear tubes that they build on the ceiling of a cave or under overhanging rocks by a river. Like spiders, they spin sticky threads that hang down from their tubes like fishing lines. Flying insects are attracted to the glowworms' lights, but when they try to fly up to them, they get tangled in the sticky threads. Then the glowworms reel in their fishing lines, gobble up their prey, and let the lines down again.

The most famous home of these glowworms is Waitomo Cave in New Zealand. Tourists from all over the world go there and see a spectacular show. They enter the Glowworm Grotto in boats that are carried in by an underground river. The tunnel entrance opens up into a huge cavern. Its ceiling looks like a starry sky. Thousands of brightly glowing larvae hang there. There are so many that their light is bright enough to read a book by. But if anyone should make a loud noise, every single light in the cave goes out. There is total darkness for a moment. Then, slowly, one by one, the tiny lights wink on again.

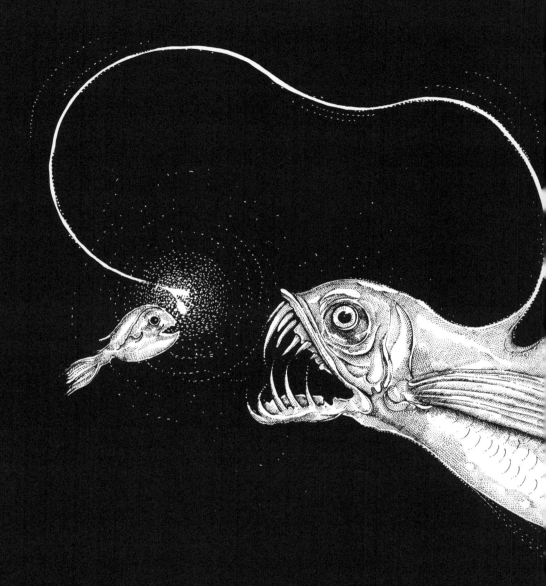

In the sea, many fish use their lights to attract their prey. One of the strangest is the *anglerfish*. It carries a light organ like a lantern dangling in front of its head on a long stalk that looks like a fishing pole. Smaller fish swim up to the light, and *snap!* they are caught in the sharp-toothed jaws of the anglerfish.

Scientists believe that the lights of dinoflagellates and other tiny glowing creatures that live in the surface waters help them to get food. When a fish swims by and stirs up the water, these bioluminescent creatures make the water glow like flashing fires. The lights may attract larger fish, which kill and eat the smaller ones. Bits of food float off into the water, and the tiny surface dwellers feast on these leftovers.

Lighted Plants and Fungi

THE uses that animals have for living lights seem fairly clear. But what good can bioluminescence be to a plant or a bacterium too small to be seen without a microscope? Some scientists think that bioluminescence first appeared long ago among some of the first forms of life that existed on earth. At that time, they believe, the earth was very different from the planet we live on now. For one thing, there wasn't any oxygen in the air. All the oxygen was trapped in rocks and water. But then plants appeared and started freeing some of this trapped oxygen. For the first life forms, oxygen was probably a very dangerous poison. Many of them died. But some found ways to use oxygen. Bioluminescence was one of these ways.

As living things found other ways to use oxygen, bioluminescence was no longer needed. But some kept it, the scientists say, because it wasn't doing any harm. And some creatures found new uses for their living lights. Scientists are not sure whether the bacteria that grow on rotting meat have any use for their living lights. (They aren't poisonous, by the way. If you eat some glowing meat or fish, it won't do you any harm.) But the glow of the luminous bacteria in the ocean may work for them by attracting fish to eat them. For most creatures, getting eaten is not a good thing. But for the bioluminescent bacteria of the ocean, it gets them a safe and comfortable home inside the fish's body.

   Bioluminescent *fungi* that live on the land, such as glowing mushrooms and the molds that grow on rotting logs, have a different use for their lights. Their glow attracts insects and other animals. They eat the fungi, but they may also pick up tiny spores that are the "seeds" for the next generation of fungi. They carry the spores around with them and help to spread the fungi to new homes.

Uses for Bioluminescence

HUMANS are finding new uses for bioluminescence all the time. The fishing industry locates schools of fish by observing the glowing trails they leave when they swim through water with luminous dinoflagellates. A special low-light camera mounted on an airplane is used to pick up the light patterns. It can pick up light 100,000 times too faint for human eyes to see. The navy is working on a similar method to track ships and submarines, and the police are using it to detect smugglers. Low-light camera mapping can also be used to study pollution in the sea by noting the changes poisonous chemicals produce in the light of bioluminescent creatures. In another kind of pollution test, samples of water are added to a tank of bioluminescent bacteria. Special strains of bacteria can be bred to be sensitive to a particular kind of chemical, such as an antibiotic drug or an insecticide.

A number of medical tests use luciferin and luciferase to detect the presence of ATP. Remember, this energy chemical is found in all living cells. Bacteria use a lot of ATP. And when people are sick with bacterial infections, some bacteria pass into their urine. (The urine of healthy people does not contain any bacteria.) So by mixing luciferin and luciferase with a sample of urine, doctors can tell if a person has a bacterial infection. This kind of test can also be used to find out if bacteria are present in drinking water and to determine if the red blood cells in the blood stored in blood banks are still good. Researchers are also working on bioluminescent tests to detect cancer cells.

The luciferin and luciferase used in such tests are obtained from the bodies of fireflies. Each summer, hundreds of children in twenty states go out into the woods and fields to work as part-time firefly catchers. The Sigma Chemical Company of St. Louis supplies them with nets and special storage cans and pays them a penny for each firefly they catch. (There is also a $10 bonus for anyone who catches more than a certain number.) Some children have sent in as many as 25,000 fireflies in a single summer. All together, the company buys about 2 million fireflies a year.

Another kind of bioluminescent test uses a chemical called *aequorin*. This chemical comes from the luminous jellyfish Aequorea, which lives in the Pacific Ocean. For a while scientists thought that this jellyfish was an exception to the rule that living lights are produced by reactions of luciferin and luciferase. They believed its light was made with just a single chemical that glows when calcium is added. (Calcium is a mineral found in living things, but especially in milk, bones, and the shells of sea animals.) When aequorin was studied in the laboratory, it was found to be a combination of luciferin and luciferase, with bits of oxygen packed inside. Calcium frees the oxygen, and then the light-producing reaction can take place. Tests for calcium using aequorin can provide an early warning of many diseases, including bone and muscle problems, heart disease, and cancer. Just a milligram of aequorin — about as much as a grain of salt — is enough for thousands of tests.

Recently researchers have isolated some of the *genes* for bioluminescence — the hereditary information that tells a cell to produce luciferin and luciferase. They have been working on a method called *recombinant DNA* for transferring these genes from one bacterium to another. They hope to put them into bacteria that are easy to grow in big tanks. Then they will be able to make luciferin and luciferase in huge amounts — tons of them — that can be put to valuable new uses.

In another study, researchers have used recombinant DNA to transfer the light-producing genes from fireflies to tobacco plants. The genes produced luciferin and luciferase and made the plants' leaves glow in the dark. These scientists are trying to find out more about how hereditary information is passed on from one generation to the next. Other researchers are studying how bioluminescence works, hoping to make lights that are as efficient as nature's own living lights.

Terms

Here are some words you may want to use in talking about living lights:

AEQUORIN. A light-producing chemical made by Aequorea, a Pacific Ocean jellyfish.

ATP. A chemical energy package found in living cells.

CALCIUM. A mineral found in milk, bones, and shells.

BIOLUMINESCENCE. Light produced by living things.

CAMOUFLAGE. Taking on colors or other disguises to match one's surroundings; for example, the bioluminescence on the bellies of fish that makes them blend into the lighted background from the ocean surface.

GENES. The chemicals of heredity, which contain the instructions that cells use for making chemicals such as luciferin and luciferase.

LARVA. The young form of an insect. It is usually rather worm-shaped and does not look like the adult.

LUCIBUFAGINS. The bad-tasting, poisonous chemicals produced by firefly larvae.

LUCIFERASE. One of the chemicals bioluminescent organisms use to make light. It reacts with luciferin.

LUCIFERIN. Another light-producing biochemical. It reacts with luciferase.

LUMINOUS. Glowing.

MIMICRY. Imitating the appearance or actions of another kind of creature; for example, Photurus females' mimicry of the light flashing of other firefly species.

OXYGEN. A chemical found in air that is necessary for burning and other energy-releasing reactions.

PREDATOR. An animal that lives by catching and eating other animals.

PREY. The animals that a predator eats.

PUPA. A stage in the life of an insect in which it is changing from a larva to an adult.

RECOMBINANT DNA. The transfer of genes from one organism to another.

SYMBIOSIS. Two different kinds of organism living together, with benefits for both; for example, luminescent bacteria and the flashlight fish.

SYNCHRONIZED FLASHING. The light-blinking pattern of Asian fireflies, in which the lights of many individual fireflies go on and off all at the same time.